RECENT ADVANCES IN PRE-TREATMENT

INTRODUCTION

Grey cloth as woven on loom is quite unattractive and contains natural components, which imparts undesirable properties to cloth and hinder the successful carrying out of subsequent dyeing, printing and finishing process. To avoid we undergo following preparatory process[1].

Grey cloth
↓
Singeing
↓
Scouring
↓
Bleaching
↓
Scouring
↓
Washing
↓
Bleached Cloth

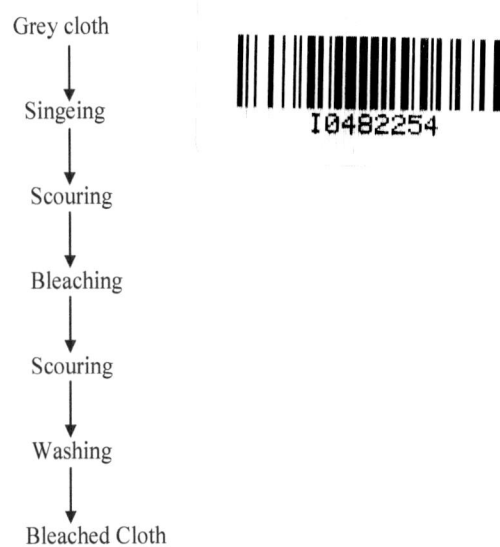

I0482254

SINGING

The process of singeing is carried out for the purpose of removing the loose hairy protruding fibers from the surface of cloth, there by giving it a smooth even and clean looking face[1].

DESIZING

In this process the size applied to the warp yarns will be removed. The major portion of size contains starch, wax and tallow. It is necessary to remove size from the cloth, otherwise the hydrophobicity of the wax and tallow hinder the subsequent dyeing and printing processes[1]

SCOURING

The main purpose of scouring is to remove fats and waxes by hot alkaline liquor containing a detergent. This includes removal of natural as well as added chemicals of essentially hydrophobic character as completely as possible and leaves the fabric in a highly absorptive without undergoing chemical or physical damage significantly. The main process occurring during scouring are saponification and emulsification[1].

BLEACHING

Bleaching include removal of natural colouring matter. The purpose of bleaching also include removal of various natural, added or acquired impurities from grey cloth as efficiently as possible with fabric in perfectly white state . It also include[1].

• Bleaching should increase absorbency
• Pure and permanent white is the criterion
• No degradation

Some traditional bleaching processes are

➢ Dilute hypochlorite (at room temperature in alkaline conditions)
➢ Hydrogen peroxide (at 80-90°c along with sodium silicate)
➢ Sodium chloride in acidic condition at boil
➢ Peroxy compounds

COMBINED PREPARATORY PROCESSES

Wet processing of the textile materials involves higher cost in each operation in the form of chemicals energy required for heating the liquor and drying the fabric. These operations consume maximum energy utilized in the fabric manufacturing and wet processing of the textile materials. Attempts have, always, been made continuously to improve the efficiency of these processes and also to combine these processes to reduce the time consumed and so the energy consumption. This topic discusses the single stage preparatory process which combines the desizing, scouring and bleaching operations into a single stage that are carried out in individual unit operations otherwise[2].

SINGLE STAGE PREPARATION

In a combined process, hydrogen peroxide or sodium hypochlorite can be used as a desizing and bleaching agent and sodium hydroxide can be used as a conventional scouring agent along with a wetting agent at pH of 10.5 – 11.5 (using soda ash). On the other hand, sodium hypochlorite can also be used for desizing and bleaching purpose with a suitable scouring agent. Various combined processes have been developed in the past using the above chemicals and also using new scourant formulations[3].

Some attempts have, also been made to review the various methods of single stage preparatory processes earlier. Preoxygen boosters were considered to be the key compound in developing a DSB process, which can, also, interact with starch and PVA sizes and to act as desize-scour boosters. Processes without peroxygen boosters were, also, developed incorporating higher strength of peroxide bur were not widely accepted. Role of peroxide activators namely tri-sodium phosphate, tetra sodium pyrophosphate, sodium hexameta phosphate, magnesium sulphate and sodium salts of EDTA were analyzed in combination of sodium silicate in the single stage preparatory process[3].

PEROXIDE BASED SINGLE STAGE PROCESS

Single stage preparatory process using hydrogen peroxide has been developed successfully in past for starch and acrylic-base sized textile materials, and, also has been used for processing various fabrics Oxidative desizing agents are active against a broad range of sizes at the same time some of them can, also, be used as a bleaching agent. In a combined process, hydrogen peroxide can be used as a conventional scouring agent along with a wetting agent. The treatments can be carried out in the stainless steel containers.

Caustic soda provides required alkalinity for scouring and activation of hydrogen peroxide. When activated, hydrogen peroxide degrades the sizing materials at lower temperatures and at higher temperature, bleaching occur along with completion of desizing. Higher alkalinity at elevated temperature produces efficient scouring action.

Silicate is used as a stabilizer and to obviate the catalytic activation of heavy metals as well as acts as a buffer. Colouring materials are decomposed into soluble products and acidic impurities in cotton fibers. These acidic products lower the pH of the bath, which in turn reduces the rate of decomposition of hydrogen peroxide and bleaching action. At the end pH reaches closer to neutral and leads substantial amount of hydrogen peroxide in un-decomposed condition on the fabric i.e. up to 21-28% and starch removal was found to be only to the extent of 82%. Using buffers like borax (pH-9), sodium bicarbonate (pH-10) and sodium hydroxide and disodium hydrogen phosphate (pH-11) were tried to counteract the acidic action of the decomposed products on hydrogen peroxide decomposition.

Buffered systems showed improved whiteness index compared to the un-buffered system since pH is maintained throughout the bleaching operation. In the un-buffered system the loss in tensile strength was observed after 55-60% of peroxide decomposition, even though there is no improvement in whiteness. The strength loss increases steadily when the pH approaches neutrality. Though hydrogen peroxide whitens the fabric in alkaline conditions (heterolytic cleavage), degradation can occur even at neutral pH (hemolytic cleavage).

$$H_2 O_2 H + HO_2 \text{ (Heterolytic Cleavage)}$$

$$H_2 O_2 OH + HO \text{ (Homolytic Cleavage)}$$

The peroxide radical under neutral conditions has a more electrophilic character and hence has the tendency to attack the cellulose rather than the coloured species. The perhydroxyl ion under alkaline conditions can attack both the cellulose and the coloured molecules. Complete decomposition of peroxide was achieved in shorter duration with higher pH i.e. pH than pH-10. Prolonging treatment beyond complete decomposition, also, resulted reduction in strength to char breakage.

The removal of star also increased with increasing pH both buffered and un-buffered conditions, i.e. addition of buffet does not affect desizing. The precipitated silica gel with high specific surface is responsible for physical trapping of perhydroxyl ion and this controls the decomposition of peroxide[4].

HYPOCHLORITE BASED SINGLE STAGE PROCESS

A self-emulsifiable solvent system of bleaching has been developed to combine the three different processes involved in the preparatory process. The system uses a high proportion of water, very low levels of solvents and hydrogen peroxide. The presence of hydrogen peroxide helps both desizing and bleaching, whereas the emulsified solvent is responsible for the scouring of cotton. Since the system involves very low solvent content, there is no need for a solvent recovery plant. Also, the operation can be carried out in plans usually available[4].

A self-emulsifiable solvent mixture was prepared consisting of a solvent (10 parts), and emulsifier and pine oil (50 parts) as a wetting agent (40 parts). Non-ionic emulsifier, perchloroethylene and pine oil combination was found to be more stable than with anionic emulsifiers. HLB values of the various agents used in the scourant mix were found to be closer to each other (pine oil 13.5, nony phenol – 13.14, and perchloroethylene – 13). Due to closer HLB values the components of the scouring agents are completely miscible and give a stable emulsion with water.

The emulsified solvent mixes with the fats and oils and thus performs the scouring action while hypochlorite acts as both bleaching agent and oxidative desizing agent. An after-treatment was given to the fabrics with cold wash, treatment with 1% solution of sodium thiosulphate to remove the residual chlorine at room temperature for 5 min. and finally again cold wash.

SODIUM CHLORITE PROCESS

Sodium chlorite affects the cotton fibres to the lesser extent, completely destroys the seed coats. The synthetic fibres can be treated with the sodium chlorite without causing any damage. The main drawback is that, at the temperatures of 80-85°C and pH of 2.5-4, the evolution of ClO_2 makes the process highly corrosive and hazardous. So different activators to make sodium chlorite effective at low temperatures like tri-ethanol amine hydrochloride (0.4%), sodium dichloro isocyanate (0.7%) formaldehyde (3 g/l) and sodium bisulphate (8 g/l) formaldehyde adduct were added to the extent of varying concentrations at temperature varied from 35-55° C for duration of 5-14 hours[4].

This process can, also, be carried out with or without buffered conditions and the recipe shown in Table – 1 is appears to have wider range in pH, time and temperature mainly due to this difference in the process conditions. In the case of un-buffered conditions, the initial pH has been kept relatively higher than the buffered conditions.

In the case of sodium chlorite hydrogen peroxide system, free radical mechanism is responsible for the bleaching action. Various free radicals created during the treatment resulted in disintegration and destruction of foreign matter in the cotton. The bleaching effect is more pronounced with peroxide than sodium chlorite in spite of the higher concentration. This is mainly due to the difference between the two oxidants with respect to decomposition in alkaline medium.

Presence of co-oxidants impedes the decomposition of each other, especially at low concentrations. Bleaching in the case of sodium chlorite is taken care mainly by the oxygen while the formation by chain reactions of radicals such as HOO and HO is responsible for bleaching action of hydrogen peroxide. Thus low levels of hydrogen peroxide (< 0.4 g/I) is responsible for poor results in terms of degree of whiteness. Also, at lower concentration of hydrogen peroxide, the decomposition products are not sufficient enough to induce activation of sodium chlorite. The mechanism under alkaline medium is as follows:

$$HOOH \longrightarrow H^+ + HOO^-$$

$$HOO^- + HOOH \longrightarrow HOO* + HO* + HO^-$$

$$HO* + H_2O_2* \longrightarrow HOO* + H_2O$$

The HO* and HOO* radicals react with the chlorite ions and, as result, a reaction chain is perpetuated as suggested by following reactions.

$$ClO_2 + HO* \longrightarrow ClO^- + H_2O$$

$$ClO_2^- + HO* \longrightarrow ClO_2^- + HO_2$$

$$2ClO_2 \longrightarrow CLO^- + ClO_3^-$$

$$ClO_2* + CLO^- \longrightarrow ClO* + ClO_2^-$$

$$ClO^* + ClO^- + HOO^- \longrightarrow Cl_2 + O_2 + HO*$$

These free radicals enhance the bleaching effect of sodium chlorite by hydrogen peroxide when used at higher concentration. Thus created free radicals seem to disintegrate the impurities sand destroy the colouring mattes of the cotton.

EFFECTIVENESS OF THE PROCESSES

In the case of the hypochlorite solvent assisted single stage preparatory process the whiteness index showed approximately linear relationship with available chlorine in sodium hypochlorite solution at various treatment times ranging from 45 min to 225 min. A whiteness index equivalent to the conventional bleaching occurs at a concentration around 6 g/I available chlorine. Tensile strength indicates a linear relationship with the available chlorine concentration. At a concentration of 4 to 6 g/l of available chlorine the strength appears to be equivalent to that obtained in the conventional method. But higher degradation results observed at higher concentration of both available chlorine and scouring agent due to severe oxidation and higher weight loss.

The decrease in the strength also takes place as the temperature of the treatment increases, due to severe action of sodium hypochlorite at a higher temperature. Wetting time decreases with increasing available chlorine concentration and no marked difference can be observed beyond 8 g/I. Better wetting time is obtained at the scouring agent concentration of 8% at a temperature of 50-55° C, which is closer to the cloud point of the non-ionic emulsifier used in the mix. Yellowing of processed material occurs on storage, similar to that occurring in conventional bleaching. It has been observed that 30-35% more NaOCI was consumed in single stage process on comparison to the conventional bleaching.

In the peroxide-alkali process. Absence of either sodium hydroxide or hydrogen peroxide in the peroxide based process, results in very low weight loss, indicating very low weight loss, indicating very low efficiency, Increase in weight loss due to scouring action and desizing occurs with increase in concentrations of peroxide and sodium hydroxide and higher weight loss is achieved at a temperature of 95°C. The whiteness index is found to be directly proportional to the concentration of sodium hydroxide, which is mainly due to the increased formation of perhydroxyl ion at higher pH, which acts as the bleaching agent and also with the processing time[5].

Also, this leads to intensified desizing and bleaching action. Increase in whiteness is obtained gradually with increase in temperature as the activation of hydrogen peroxide increases with temperature. Wettability improves with increasing concentration of sodium hydroxide and the lowest wetting time was observed at 95°C.

In peroxide-solvent assisted scouring agent based process substantial weight loss, strength loss a lowering of wetting time occurs up to a concentration of 4% of scouring agent. Also a very small change in these parameters was observed at higher concentration levels. For a given temperature and scouring agent concentration a rapid increase in weight loss, a steep decrease tensile strength and wetting time are observed with respect to an increase in peroxide concentration upto2%.

Apparent linear relation exists between a) weight loss and tensile strength. B) Wetting time and weight and loss and c) strength and wetting time. But significant correlation was observed only in the case of wetting time and weight loss. No correlation was observed between weight loss, tensile strength and wetting time.

Weight loss increases gradually with increase in concentration of bother peroxide and scouring agent and similar trend occurs in the case of wetting time also. Time and temperature also directly affect the weight loss. There is a decrease in tensile strength with respect to the peroxide concentration at all concentrations of scouring agents. However, at lower concentrations, most of the fall in strength occurs below 2% peroxide[2].

Whiteness index show a linear relationship with the concentration of hydrogen peroxide and other factors like scouring agent concentrations, time and temperature had a little effect. Under identical conditions, acrylic, sized fabrics showed superior results than scratch sized fabrics in terms of whiteness and wettability. The poor results in the case of starch sized samples can be attributed to higher add-on levels of size and incomplete removal of tallow from the fabric during desizing.

In the case of combined process using sodium chlorite-hydrogen peroxide, the whiteness increases with increasing time and increasing pH (8 to 10). This seems to be efficient for 100% cotton, sized with starch and P/C blends (50/50). Relatively a lower degree of whiteness is achieved in the case of blended fabric. The weight loss occurs to the extent of 8.01 to 8.55% with the tensile strength 81-86% of the grey fabric strength.

In the single stage preparatory process, an attempt has been made to improve the efficiency by carrying out the process using microwave with the energy of 600W for duration of 30s to 30 min. The process involves steaming of the fabric with the simultaneous exposure to the microwave energy.

Table-1 Recommended Recipes for Single Stage Preparatory Process[2].

No.	Process	Recipe
1.	Peroxide-alkali	H_2O_2-1.0 g/I, NaOH – 10g/I Na_2 Sio3-5g/I, Wetting Agent –1g/I Temperature – 95°C Time –120 min PH – 10.5 11.5 using soda ash
2.	Peroxide Solvent assisted Scourant Starch sized fabrics	Scoring agent – 4% Peroxide – 2% Sodium silicate – 1% Wetting agent – 0.1% Temperature – 95°C,m Time 180 min pH –10
3.	Peroxide Solvent assisted scourant Acrylic-base size	Scouring agent –4% Peroxide –1% Sodium silicate –1% Wetting Agent 0.1% Temperature –90°C, Time –180min pH -10
4.	Peroxide-Solvent assisted scourant	Scouring agent – 4% owf. H_2O_2 –1% Na_2 Sio3 – 2% Na_3 PO_4 – 2% Or tetra sodium pyrophosphate PH –11 Temperature –60 min Time –12-16/24 hours at 40°C
5.	Hypochlorite- solvent assisted scourant	Solvent based scouring agent –2% NaOCI-6 g/I (av.CI2) Temperature –40°C Time –180 min, M:1 –1:20

		PH –11
6.	Sodium chlorite- Peroxide Process	$NaCIO_2$ –3 g/I, Na2 HPO4 –10g/1 NI Wetting Agent-2 g/I, pH –10 Time –90 min, Temperature –95°C M:1 1:20
7.	Sodium Chlorite Solvent assisted scourant	Sodium Chlorite –0.8 –1.95%, Scouring Agent-2% Activators –As said above in the text Temperature –30-55C, Time –5.14 hours pH –4.6 (buffered) & 8.9 (un-buffered)
8.	Peroxide-Alkali Process	Sodium Hydroxide – 20 g/I Hydrogen Peroxide –30-40 m/I, Peroxidesulphate-5 g/I Sodium Silicate –20 m/I, Surfactant 10 g/I Stabilizer –5 m/I
9.	Flash Steam Process	Ciba Tinoclarite FS – 15-30 ml/kg NaOH (100%) – 30-50 g/kg, H_2O_2 (35%)-49-90 ml/kg

In another novel approach steam procedure has been us combine all the prepare processes into a single step helps to complete the fact preparation within 2-4 minutes, applying the solution, consisting alkali, hydrogen peroxide and auxiliary to the fabric, it is steam with saturated steam followed by wash.

COMBINED DESIZING SCOURING-BLEACHING MERCERISATION PROCESS

Though it has been considered to difficult to combine all the processes into a single process due to the difference in the temperature involved, an attempt has been mach in the recent past to perform the unit processes in a single stage using curing operation after impregnation with the various ingredients. Cotton fabrics. After impregnation with a mixed solution containing sodium hydroxide hydrogen peroxide and trichloroethylene for 3 minute at different temperatures and fabrics are taken for curing at $120^{\circ}C$, followed by neutralization and washing. In this case, the strength retention, whiteness index, dye ability are mainly controlled by concentration of sodium hydroxide, dipping time, temperature and curing duration. But no further literature is available in this regard.

RECIPE RECOMMENDED FOR SINGLE STAGE PREPARATORY

From the various literatures available in the area of single stage preparation, different recipes can be observed for combined DSB process. Following table gives the various recipes developed for the single stage preparatory process.

CONCLUSION

It is possible to use sodium hypochlorite/hydrogen peroxide/sodium chlorite along with a self emulsifiable solvent assisted scouring system for a single stage preparatory process. This single stage process results in 5-17.5% savings in energy and proved to be a cost can be saved compared to the conventional process methods. Attempts have also been made to use hydrogen peroxide along with alkali to combine the various processes involved in the fabric preparation. Combining mercerization in this single stage preparatory process still remains unexplored.

Pseudo single-bath process for alkali treatment and bleaching of jute at ambient temperature

Pseudo single-bath process of alkali treatment and bleaching was carried out using the following four sequences:

Grey jute fibres were dipped in sodium hydroxide solution (25 gpl) at 1:3 material to-liquor ratio, kept as such for 1 h, squeezed and taken out. The alkali-treated fibres were then dipped in a water bath (M:L ratio, 1:3) for 15 min. Then all other ingredients, except sodium hydroxide, were added to the bath and ambient temperature bleaching was carried out in the usual way[6].

Grey jute fibres were dipped in sodium hydroxide solution (50 gpl) at 1:3 material-to-liquor ratio, kept as such for 1h, squeezed and taken out. The alkali-treated fibres were then dipped in a water bath (M:L ratio, 1:3) for 15 min. Then all other ingredients, except sodium hydroxide were added to the bath and ambient temperature bleaching was carried out in the usual way.

Grey jute fibres were dipped in sodium hydroxide solution (50 gpl) at 1:5 material-to-liquor ratios, kept as such for 1 h, squeezed and taken out. The alkali-treated fibres were dipped in a water bath (M: L ratio, 1:3) for 15 min. The alkali coming out to the water was then titrated and the concentration of sodium hydroxide was found to be 21.6 gpl. Hence, 3.4 gpl sodium hydroxide was added to the bath along with other bleaching ingredients for ambient temperature bleaching. The fibres were then squeezed to give 100% wet pick up and kept in a plastic bag for 2 h. Finally, they were washed, soured and dried.

Grey jute fibres were dipped in sodium hydroxide solution (60 gpl) at 1:3 material to-liqour ratio, kept as such for 1h, squeezed and taken out. The alkali-treated fibres were dipped in a water bath (M:L ratio, 1:3) for 15 min. The alkali coming out to the water was then titrated and the concentration of sodium hydroxide was found to be 25 gpl. Hence, other bleaching ingredients except sodium hydroxide were added to the bath for ambient temperature bleaching. The fibres were then squeezed[6].

COMBINED DESIZING SCOURING AND BLEACHING OF COTTON USING OZONE

The increasing demand for the conversation of natural resources and environmental protection has forced the researches to look for the processes which can be carried out at a low temperature using small amount of water in a short duration without the use of harmful chemicals such as hypochlorite[7].

To fulfill these objectives, attempts are now being made to bleach cotton using ozone, ozone-UV radiation, peracetic acid and enzymes. Ozone being a powerful oxidizing agent can bleach cellulose much faster than hydrogen peroxide and hypochlorite. Today, it is being used, to a large extent as a substitute for chlorine in pulp bleaching.

However, in cotton bleaching it is still at the infant stage. In 1995, two patents have been awarded for bleaching of cotton with ozone. These patents claim a continuous process using low-temperature plasma for desizing and scouring and ozone in the presence of UV for bleaching.

In the present work, an attempt has been made to combine all the three grey preparatory processes, such as desizing, scouring and bleaching. Here, ozone is used to desize, dewax and decolour the grey cotton fabric. To improve the fabric properties two stage bleaching is suggested wherein the fabric is treated with ozone followed by hydrogen peroxide[7].

APPARATUS

The equipment used for the bleaching with ozone has three components: the ozone generator, the applicator and the ozone destroyer. The ozone generator of 8g/h capacity was supplied by Ozone Tek Ltd, India. The input for the generator is oxygen from a pressurized cylinder. The generator supplies the required concentration of ozone-oxygen mixture to the applicator. The applicator is a glass cylindrical tube with a diffuser at the bottom.

The gas mixture is pumped continuously as the required flow rate into the applicator through the diffuser. The wet fabric samples are suitably placed in the applicator for the required time. The spent mixture is passed through the ozone destruction unit which has a heating element, where the ozone is completely converted into oxygen before being released into the atmosphere. An ozone analyzer is mounted on the applicator to measure ozone concentration both at the entry and the exit of the applicator[7].

PROCEDURE

Wet grey cotton fabrics having 24% moisture content (wet pickup) were placed in the application chamber and exposed to 100 g/m concentration of ozone oxygen gas mixture at pH 5 using acetic acid for a specified time (1-7 min) at room temperature (about 30 C) for the combined desizing, scouring and bleaching. The gas mixture flow was maintained at a constant rate of 0.5 liter/min for the all experiments.

Two-stage bleaching of grey cotton fabrics was first carried out with ozone and then with hydrogen peroxide. The wet grey fabrics with –24% moisture content were first bleached with ozone (100 g/m) at pH 5 for 1, 2 and 3 min and washed. These fabrics were then bleached with hydrogen peroxide for 45.30 and 15 min respective y using standard procedure. All the bleached samples were then hot soap washed at 80 C, cold washed, dried and finally conditioned before testing for the properties[7].

Grey fabric was also acid desized, alkali scoured and bleached with hydrogen peroxide and calcium hypochlorite using standard procedures.

CONCLUSIONS

The above study shows that grey preparation with ozone can be completed in one or two minute. This process has additional advantages such as savings in thermal energy, water and chemicals. For an acceptable degree of whiteness which is sufficient for dyeing medium and dark shades with reactive dyes, the quality of the fabrics processed with ozone is comparable with that of the fabrics processed by conventional method of desizing scouring and hydrogen peroxide bleaching.

For high degree of whiteness, undoubtedly the quality of fabrics bleached with hydrogen peroxide is found to be superior to that of the fabrics bleached with ozone and calcium hypochlorite. However, it can be said that there is a scope to improve the quality of ozonized fabrics. A two-stage bleaching of grey fabric with ozone followed by hydrogen peroxide is recommended for achieving a high degree of whiteness with minimum cellulose modification, which may be suitable for dyeing pastel shade[7].

ENVIRONMENTALLY FRIENDLY PROCESSES

Alkaline scouring is a commercially well established and efficient process. The drawback of this process, however is the use of large amounts of caustic (up to 80 kg/ton fiber material) as well as large quantities of rinse water once the scouring process is completed while the rinse water could be reused, the scouring efficient with its high BOD, high COD, and alkalinity presents a major problem Increasingly stringent effluent regulations have thus generated interest in procedures that are nontoxic and environmentally benign[8].

Pectinase

Pectinic components as found in raw cotton mainly consist of neutral and acidic hetero polysaccharides with different molecular weights and degrees of esterification. They are comprised of three major groups pectic substances, which are composed of partially or completely esterified polygalacturonic acids, protopectins, which are water- insoluble pectins chemically or physically associated with other cell wall components and pectins acids, which are formed from non-esterifed polygalacturonic acids. The lower the molecular weight and the higher the degree of esterification the better the water solubility of these compounds, if occurring in water-insoluble salt form pectinic compounds are associated with calcium, magnesium, or iron ions[8].

Pectinases capable of hydrolyzing pectinic substances are generally enzyme complexes containing esterases and depolymerases with random or terminal activities. Their activity optimum usually lies in the slightly acidic region an the temperatures around 50C. With the conventional caustic scouring procedure it is not possible to completely remove pectins especially if they are associated with calcium.

Cellulase

Cellulases enhance the effect of pectinase to a certain extent and add softness to the cotton fabric. They often accompany pectinases in small amounts. If used or scouring cellulases hydrolyze cotton cellulose lifting off non-cellulosic impurities in the course of the reaction. Typical fiber damage caused by cellulase cannot be excluded[8].

Lipase

Cotton waxes consist of various hydrocarbons, fatty alcohols and acids, and their respective esters. These fats and waxes are the major reason for the hydrophobic nature of the unscoured cotton fiber during the conventional caustic scouring process, sodium hydroxide forms a soap with the fatty acids, which is turn removes the waxes and other compounds.

Another option is scouring with organic solvents during which these hydrophobic components are extracted Scouring in presence of a non-specific lipase can ease the removal of fats and waxes. Many lipase enzymes are capable of non specifically cleaving various natural lipids and oils, generating glycerides, glycerol, and short-chained fatty acids[8].

Xylanase

Hemicelluloses are a heterogeneous group of polymeric compounds based on B-glucancs, mannose, rhamnose, and other substituted hexoses as well as xylose and arabinose. A portion of these hemiscelluloses is water soluble, other are bound to microfibrils via hydrogen bonding. Xylanases proved to be very useful in combination with cellulases for cotton finishing these enzymes are capable of cleaving B-14, linked xylosyl sequences from hemicelluloses[8].

Peracetic acid bleaching of cotton.

Since peracetic acid is a biodegradable chemical, bleaching with it is an environmentally friendly process, peracids and hydrogen peroxide have along been the principle bleaching agents for domestic laundry, but despite vigorous research, the mechanism of peroxygen bleaching remains a mystery. An understanding of this mechanism is essential for proposing peracetic acid bleaching as an alternative to hypochlorite bleaching.

They have examined five variables of peracetic acid bleaching, using statistically designed experiments the amount of hydrogen peroxide, amount of acetic anhydride, treatment temperature, time and pH. The results of our investigation show that the major factor determining whiteness in treatment temperature, the second factor is treatment time, the third factor is acetic anhydride concentration, the fourth factor is hydrogen peroxide concentration, and the fifth factor is pH[9].

Although whiteness of the cotton fabric is not significantly influenced by pH, the other four factors have a significant effect on it. We have studied a very narrow pH range, and the bleaching mechanism is approximately the same for that whole range. Thus, the effect of pH on whiteness seems to be the smallest one. If we had studied a wide pH range, its effect would be quite different from this situation. We preferred to study a narrow pH range to eliminate the major role of pH. The results of our investigation also show that all of the factors do not have a significant effect on bursting strength.

The major factor determining water absorbency is treatment temperature, the second factor is pH, the third factor is acetic anhydride concentration, the fourth factor is hydrogen peroxide concentration, and the fifth factor is treatment time. The water absorbency of cotton fabric is significantly influenced by treatment temperature and pH, while the other three factors do not have a significant effect on it.

Peracetic acid is a biodegradable bleaching agent, and not using excessive amounts of chemicals in the process is very important for the ecology. Not using a higher temperature and treatment time is also important in terms of economy. We have optimized with five variables, keeping the other parameters constant, and three responses. The aim of the optimization is not to use parameters more than necessary to obtain the desired response values .

To obtain an optimized bleaching recipe, considering all responses the % ideal reflectance value at 460 mm is 80, the ideal water absorbency time is 5 seconds, and the ideal bursting strength is 12 kg/cm. To optimize the recipe the % reflectance value at 460 mm is over 75, the water absorbency time of the sample is lower than 60 seconds, and the bursting strength value is over 9 kg/cm2. We have studied 59,049 points, and determined the estimated optimized process parameters with a computer. The optimized level of hydrogen peroxide is 2 ml/l, acetic anhydride is 3 ml/l, temperature is 45 C, and treatment time is 52.5 minutes of the pH is 7.25. The results show that this experience design can be used to optimize peracetic acid bleaching[9].

Water Preservation in the Textiles Industry[24]

Water is an increasingly precious resources, yet the textile industry, a major use of water, takes little are to preserve what can no longer considered an infinite resource said that the cause of a third world war would water ability and rights. Whilst major efforts are being made within the textile industry to reduce in order to increase profitability gain a competitive edge, such efforts are mainly cent red upon manpower reduction or increase in process and machinery efficiency. The cost of water and the associated costs of heating and disposal are generally brushed aside as 'fixed costs.

The textile industry cannot operate without water. Water is vital to the processing of textiles. It is therefore essential that the industry make every effort to minimize the volume used, on both commercial and environmental grounds. If this can be achieved there will be also the associated reduction in the energy required for heating water. By carefully selecting chemical auxiliaries that allow process solutions to minimize the use of water the chemical industry can help solve this major problem.

The pretreatment process has traditionally been carried out in a logical and systematic way and impurities presented to the pretreatment technician are thoroughly scrutinized. A bale of grey woven cotton fabric may contain up to 20% of its total weight in impurities, and these must be removed in sequence.

Problems with the Pretreatment Process

Cotton is a natural, vegetable fibre and will have a variable composition dictated by the local soil and growing conditions. The main natural impurities are cotton waxes and metals that are held together by pectin and form the primary wall of the fibre. Although only 1% of the fibre's diameter, it contains 90% of the natural impurities. This primary wall, or cuticle, is hydrophobic and must be removed to allow for successful processing. Depending on the region and crop year, the metal content can vary.

This presents the pretreatment process with special challenges to consistency. In that there can be adverse interactions between the metals and the hydrogen peroxide. In addition to natural variations found in cotton, the mill will add other processing aids that facilitate the manufacture of yarn and cloth. Processing aids include sizes and lubricants, added during spinning, weaving or knitting stages. typically, about 25 liters of water will be required for every kilo of cloth being processed. In a pretreatment process of this type, the singeing, cooling drying stages cannot be rationalized Desized, scouring and bleaching, however, can be rationalized initial to two stages, showing some savings water, but ultimately to a single stage process.

Single Stage Process

Single-stage pretreatment has been shown to significantly reduce water steam and processing times in the mill. Water usage can be reduced by almost 60% by adopting a Single Stage Process. Single Stage: Pad Steam incorporates Mechanical De-sizing 10.5 liters. The rationalization to both Single and Two Stage options has been made possible by using specialist chemicals auxiliaries, introduced by Allchem International Limited to enable the desizing, scouring and bleaching to completed consistently.

Further Development showed the fabric was pre-washed; the reserve was an improvement in pretreatment consistency. Pre-washing has led number of purpose-built, single steamer pretreatment units being erected and successfully operated many years in the U.K. Such units operate efficiently, showing the associated increased productivity water savings. The single-stage process has been taken a step further using specialist vacuum technology and encompasses further desirable environmental opportunities. The use of the vacuum technology will result in an additional reduction in water consumption but always leaves a small amount of size prior to the chemical impregnation.

RECENT DEVELOPMENTS

Desizing[23]

Cold Pad Batch H_2O_2 Desizing

The cold pad batch technique was best suited for the smaller processor or circumstances where the fabric was not suitable for continuous processing due to creasing etc. (Table I).

TABLE I. Cold Pad Batch H_2O_2 Desizing

Batch Composition	Amount (g/L)
H_2O_2 35%	20-35
NaOH 100%	30-45
Solopol APS	4-7
Cinscour VIC	4-8

Pickup 100%, batch 8-24 hrs.

Pad Steam $H_2 O_2$ Desizing

The pad steam technique was easily added to existing scouring equipment the availability of hydrogen peroxide on the pretreatment line meant minimum modification was needed to adapt the process.   The results obtained were very good. The pad steam process gave completed removal of all size types after the desizing stage. In the cold pad batch technique, desizing was completed after the bleaching process.

TABLE II. Pad Steam H$_2$O$_2$ Desizing

Batch Composition	Amount (g/L)
H$_2$O$_2$ 35%	5-9
NaOH 100%	35-55
Solopol APS	1-2
Cinscour VIC	5-8

Pickup 100%, steam 5-15 min.

Cold Pad Batch H$_2$O$_2$ pretreatment

The lack of thermal energy in the cold pad batch H$_2$O$_2$ pretreatment process meant the chemical intensity needed to be increased to compensate Development work showed that in the case of cold pad batch simply increasing the concentration of the formulation was not cost effective. To improve the quality obtained, particularly the degree of desizing, it was beneficial to add persulfate to the formulation. In the formulation listed this is contained in Cinscour CSP. Development of the pad steam single stage technique followed. Based upon data collected from the cold pad batch work, certain criteria had been established that enabled the first work trials to be undertaken. After refinements the technique and formulation was firmly established.

TABLE III. Cold Pad Batch H2O2 Pretreatment

Batch Composition	Amount (g/L)
H_2O_2 35%	60-80
NaOH 100%	20-30
Solopol APS	15-20
Cinscour CSP	9-12

Pickup 100%, batch 16-24 hrs.

Mechanical Desizing

A new phase of the single stage process began five years ago when the mechanical desizing process was introduced. This is the addition of a vacuum slot after the prewash section of the pretreatment machine. The fabric is wet out/washed in hot water for 20-60 sec after which it passes over a vacuum slot. The majority of the size is removed at this point. Removal between 89-99% of the size is seen. The size type does not particularly determine the amount of size removed. It is influenced more by the fabric construction basically the more dense the fabric, the lower the level or removal. The fabric is subsequently processed using the single stage formulation.

Conclusion

During this development work, one factor stood out as a dominant influence on the quality of pretreatment achievable. The products used in the sizing process massively influenced the outcome of the process. Using the conventionally accepted procedure based on enzyme desized, alkali scour, and bleach, certain combinations of size products were not removed at the desized stage and caused problems further into the process. For example, if a fabric was sized purely with PVOH, enzyme desizing would have minimal effect on its removal, meaning that a considerable amount would pass into the alkali scour. It would pass into the alkali scour. It would be further saponified leading to difficulties in its removal.

Another example would be a fabric sized with a native starch. If the fabric was passed through a size reclaim process designed for synthetic sizes, the starch would not be removed at that stage or in a subsequent alkali scour. It would pass through to the bleach stage and lead to poor bleaching results.

Developments that have taken place in recent times in scouring and bleaching of textile materials are reviewed in the following.

Cotton Scouring:-

A number of enzymes for scouring of cotton fabrics have been developed. Thus some workers have studied the effect of sonication (with ultrasound of 1.4KW at 16 and 20KHZ frequencies) on cotton preparation with alkaline pectinase and acidic pectinase, Viscozyme L of Novo Biochem, inc. Their research at the laboratory level has shown that introducing ultrasonic energy improves pectinase activity, but does not decrease the tensile strength of the fabric[10].

Further, sonification of pectinase processing solutions does not impair the complex structures of the enzymes molecules, but significantly improves their performance; alkaline pectinase has proved to be a more efficient agent for biopreparation of grey cotton fabric than acidic pectinase, resulting in better wettability and whiteness. The combination of pectinase discoursing with desizing and after washing ensures sufficient fabric wettability and adequate uniformity.

The results are comparable to or better than those for fabric after the traditional alkaline scouring. Introducing ultrasonic energy into the reaction chamber during enzyme treatment of cotton fabric could help overcome the major disadvantage of pectinase scouring – longer processing time, compared to conventional alkaline scouring.

In another process, had earlier found that proteases are effective scouring agents for boiling water- pretreated cotton fabrics. Later they have applied the enzymes directly on grey cotton fabrics. Direct reactions with three proteases-trypsin, chymotrypsin, and subtilisin on these fabrics and found that all the three enzymes improve fabric wettability to as level similar to alkaline scouring, under mild conditions (45°C-55°C at pH 7)[11].

The reaction conditions required to achieve optimal fabric wettability (Cos 0 > 0.6) are 5 gpl and 45°C for subtilisin, chymotrypsinis effective under several reaction conditions, i.e., 1 gpl at 55°C , 2 gpl at 45°C; 5 gpl at 35°C. Most reactions takes place in 30 min and room temperature water rinses replace the post-reaction buffer rinses (of the earlier study). Compre4d to the earlier process, direct protease scouring requires increased concentration (subtilisin), higher temperature (subtilisin and chymotrypsin), or longer time (trypsin and chymotrypsin) to achieve similar water wettability and absorption properties.

The most direct outcome of protease scouring of grey cotton fabrics, in comparison with the earlier process and alkaline scouring, are the resulting fabric characteristics , i.e., less lateral shrinkage, no change in surface friction, easier to shear, and more resilient to compression and bending.

Cotton bleaching:-

Someone have studied silicate and non-silicate – stabilized hydrogen peroxide bleaching of cotton fabrics. Both single-stage, and two stage bleaching procedures have been studied to optimize the peroxide concentration in the bleach bath. It was found that the optimum concentration was 3% (owf) for the two stage process as well as for the single stage process, like scourex or modern bleaching process shows that the non-silicate stabilizers can be used in the place of silicate without an adverse effect on the uniformity and levelness of the subsequent dying (with three HE-type and one H-type tryazinyl reactive dyes) [12].

The bending length data shows that the stiffness of the sample bleached in nonsilicate – stabilized system is lower than that in silicate-stabilised peroxide bleach bath, indicating that the problem of silicate deposits on the fabrics can be avoided. Some industrial trials have been concluded, the results of which support the laboratory results.

Some one have designed close system bleaching apparatus to determine the kinetics and the effects of various parameters on the alkaline hydrogen peroxide bleaching of textiles cellulosic fibers. It was confirmed that per-hydroxyl anion (HOO⁻) is the primary bleaching moiety in alkaline hydrogen peroxide systems. The use of the apparatus in the measurement of the fabric colour, waste oxygen and the subsequent calculation of perhydroxyl anion and unused molecular hydrogen peroxide confirmed that pH and titration of free hydrogen peroxide in the alkaline bleaching system are not good indicators of bleaching mechanisms.

The role cellulose itself in the chemical engineering systems was determined. The rate of bleaching of cotton fabric was shown to be a first order reaction in the concentration of perhydroxyl anion at $60^{\circ}C$ and $90^{\circ}C$. An activation energy 17kcal/mole was estimated. Decomposition of hydrogen peroxide was found to follow second order kinetics.

OZONE BLEACHING:-

In an earlier study some of them showed that bleaching of wet grey cotton fabric can be successfully carried out using ozone to get an acceptable degree of brightness (ready for dyeing). The process resulted minimum mechanical and chemical damaged to cellulose. The study has been extended to the roles played by moistures and pH during ozone bleaching and design features of the ozone application chambers. The bleached fabric were dyed with 7 reactive dyes and 2 vat dyes, and finished with DMDHEU cross linking agents in the presence of magnesium chloride catalyst.

The properties of this fabric, whiteness index, aldehyde groups, carboxylic groups (for a bleached samples), colour difference, crease recovery angle, nitrogen content and strength loss (after cross linking) with those of conventionally hydrogen peroxide- bleached (and dyed and crosslinked) fabrics. Considerable chemical damaged (aldehyde group formation) occurred in ozone bleaching and colour difference of ozone bleached and peroxide bleached sample exceeded 3.0, indicating lower dye ability of the former.

Maximum whiteness was obtained when the moisture content of the fabric prior to ozone bleaching was about 24%. There was margin decrease in whiteness over the pH range 2-9, and considerable pH 10 and 12. the strength loss (about 5 to 6%) in the pH range 2-6, reached up to 20% at pH 12. it is claimed that bleaching with high concentration of ozone (100gm/cu.m) in shortest possible treatment time at pH 7, with moisture content of 24% would be superior to all the conventional methods of bleaching grey cotton fabrics.

Enzymatic bleaching:-

Enzymatic bleaching of cotton fabric with glucose oxidase has been reported. in the study, grey cotton fabric was desized with amylo-glucosidase and combined with bioscouring with two different kinds of pectinase in the presence and absence of cellulose. Then a process has been established that allows one to combine enzymatic desizing, bioscouring and enzymatic bleaching with glucose oxidase, using the hydrolysis of glucose for the production of hydrogen peroxide (a known and widely used bleaching agent)., and then re-using the treatment bath. Bioscouring, if performed in the presence of cellulose, yields a materials with overall good properties, but with reduced tensile strength, indicating the well known degradation pattern of cellulose[13].

If cellulose is replaced by small additions of a nonionic surfactant . comparable properties are observed without the problematic strength reduction. One-bath enzymatic desizing/ scouring/ bleaching is feasible when sufficient amylglucosidase for glucose production is present. Whiteness indices of the enzymatically bleached goods are close to those of fabrics bleached conventionally with hydrogen peroxide. With lower energy costs and less rinse water usage, the process presents an economically interesting alternative for the textile preparation technology.

Dyeing of bleached cotton fabrics with two reactive dyes (Evercion blue HE-GN, CI Reactive blue 198; and Everzol Brilliant orange 3R, CI Reactive orange16) in a peroxide bleach bath after destroying the residual hydrogen peroxide with the enzyme (catalase) has been reported. The conventional rinsing after bleaching to remove the residual hydrogen peroxide, which is harmful to some dyes, either in solution or on the fabric, was replaced by enzymatic cleansing using catalase[14].

Bleaching baths in the present study, were treated prior to dyeing , with catalase to convert the residual hydrogen peroxide to harmless gaseous oxygen and water. There are some limitations. Poor stability of the enzyme at high temperature and pH, the influence of the bleach both composition on enzyme efficiency and on the dye uptake. To prevent the deactivation of the enzyme, the bleach bath was cooled before the addition of the enzyme and the alkali, neutralized. Though the peroxide was completely destroyed by the enzyme, the bleach bath formulation caused unacceptable changes in the dyed fabrics.

By varying dye , salt, alkali, and enzyme concentrations, it was shown that increasing the amounts of these components to certain levels, the colour difference could be reduced significantly and set to the defined colour difference. Selecting the optimum proportion between each of these parameters could be the key for the successful dyeing in the bleach bath to realize considerable water and energy conservation, avoiding the extensive washing cycle after bleaching.

WOOL BLEACHING:-

Some one have analyzed the conditions of continuous bleaching of Australian and south African wool. The hydrogen peroxide concentration and pH of the bleaching solution were chosen as the parameters, since the other process variables, i.e., the time of contact with the solution, percent impregnation, drying time, and temperature were applied under conditions similar to those used in the industrial processes. The whiteness index rises almost in direct proportion to peroxide concentration and the pH of the solution has little influence on the degree of whiteness obtained. Under the experimental conditions, carbonized and bleached wool does not undergo significant chemical alterations during the bleaching process, evaluated in accordance with their alkali solubility and cystic acid content.

In another study on wool bleaching some of them have carried out surface modification and bleaching of pigmented wool. Naturally pigmented karakul wool was subjected to a surface modification system of chlorination, using Basolan DC (dichloroisocynuric acid in saturated salt solution) followed by dechlorination with sodium bisulphite treatment. The fibers were then catalytically bleached by a two-bath process (mordanting with ferrous sulphate and bleaching with hydrogen peroxide in separate baths). The fibers, thus treated were examined for their structure and properties[15].

It was found that the surface modification removed most of the scales, resulting in only a small degree of damage to the tensile properties of the fibers. The bleaching increased the fiber whiteness and did not cause apparent changes in the surface morphology of the wool. Even though the fiber epicuticle remained almost intact after bleaching fiber strength suffered further losses. However, the modified and bleached karakul wool still has greater strength than white cashmere. After bleaching, the felting propensity of karakul wool improved slightly, and its dye uptake (with CI Acid Red) decreased.

In a similar study, they studied the effects of ferrous mordanting on bleaching of pigmented wool, has been improved. The serious problems of discoloration and excessive damage of bleached fibers associated with the deposition of iron during mordant bleaching process are resolved. All mordanting parameters are critical for the successful bleaching of pigmented fibres. When used at a specific temperature and a specific time during mordanting ,a reagent (sodium bisulphite) or an acid (acetic or citric acid), capable of stabilizing the mordant bath during mordanting , produces the most complete reaction of melanin with the iron, while not allowing the iron to be deposited on the fibre[16].

Thus, in the subsequent bleaching process, maximum whiteness is achieved, with less damage in terms of the mordant. Sodium sulphite can be used as an auxiliary, i.e. at the initial pH of the bath, and acetic acid or citric acid can also be applied during mordanting without causing discoloration and excessive damage. The optimum conditions for mordanting are: 20°C, for 15 min, when sodium sulphite is used , and 60°C for 15 min with acetic or citric acid stabilizer. The improved process is superior to conventional processes, with no need for after treatment with a reducing agent.

Another report has been shown for a process of rapid bleaching of wool with reducing agents. In this study the effectiveness of using different reducing agents with or without sodium lauryl sulphate (SLS) has been compared. The bleaching was done for 1 and 2 hr. the former was found to be less effective when using TUD(thioureadioxide) but not when not using Blankit IN , sodium bisulphite and sodium hydrosulphite. The bleaching time did not affect the amount of cystein in wool when using Blankit IN or sodium hydrosulphite, but had significant effect when using other two reagents. Bleaching for 1hr in presence of SLS yielded better results than bleaching for 2 hr in absence of SLS. Regardless of reducing agent. Chemical aggregation was less when using TUD, sodium bisulphite or sodium hydrosulphite.

Pyrophosphate/ sodium oxalate) and acidic (prestogen W) media using hydrogen peroxide has been studied to access the degree of whiteness attained the severity of chemicals attack on the fibre to achieve a pre-determined whiteness. It was found that bleaching under alkaline conditions led to much white wool than bleaching in an acidic medium; however, the former process produced a much more intense chemical attack than the latter, reduces from the higher contents of cysteic acid and alkali solubility's[17].

Thus, an excellent linear correlation between cysteic acid concentrate and alkali solubility was obtained, the points for alkaline process failing on a steeper straight line than those for the other process, indicating that for a given cysteic acid content, wool bleached in an acidic medium has a higher alkali solubility than that bleached in an alkaline medium. It is suggested that bleaching leads to the formation of intermediate oxidation products of the disulphate bond. As these are unstable in an alkaline medium, they will be converted into cysteic acid during alkali solubility test and hence give rise to higher alkali solubility values those that would correspond to the cysteic acid content substrate tested.

Bleach activators:

Has described certain bleach activators. The leading bleach activator in Europe for the last twenty years has been tetraacetyl ethylene diamine (TAED), which produces peracetic acid in situ by reacting with a peroxy , etc. it is cost effective , environment friendly and provides effective bleaching at as low a temperatures as 40°C. The search for alternatives to TAED has been going on since it was first launched on the detergent market in 1979. At Warwick int., they have tested 1,000 bleach activators and have assessed them for their wash performance, environmental effects, costs, and ease of synthesis. To illustrate this work, the results of the bleach activators, 2-substituted3, 1-benzoxazinones are presented.

The effect of bleach activator in detergents on the fastness of dyed cotton fabrics has been studied. It is pointed out that the incorporation of a bleach activator into a detergent formulation (along with a peroxy oxidant) allows domestic laundering to be conducted at low temperatures using lower volumes of water. However, the generated bleaching species can produce oxidative bleach fading the dyed article during its laundering life time. A diagnostic test was devised, which indicated the dye that exhibits excessive fading, thereby allowing all participating in textiles processing chain to produce high quality colored articles. This test has already become a British standard-BA 1006: UK-TO colour fastness to domestic laundering oxidative bleach response and is expected to be adopted by ISO in the near future.

They have studied the colour fastness of optionally after treated with direct dyes on cotton during washing in an activated oxygen bleach-containing detergent, using a previously proposed UK-TO spectrophotometric test. Eleven direct dyes, representing different but important types of chromophores, seven dye fixing agents, and four wash liquors, phosphate-free detergent, peracetic acid formed in situ form sodium perborate and TAED were used in the stud. It was found that seven of the eleven dyes studied gave instances of fading (colour change) due to bleeding and/or oxidative effect of peracetic acid, present in the was liquor, and that after treatment with dye-fixing agent, generally reduces fading. However, in some cases of after treatment, the fading of the dyes increased, to explain which, further research is needed. The UK-TO test method successfully predicts which combination of direct dyes and after treatment will exhibit unacceptably high levels of fading during repeated washing in an activated oxidative bleach-containing detergent[18, 19].

Some work has been reported on this topic. Thus, they have conducted a two-part study on the colour fastness of optionally after treatment acid leveling and milling dyes in knitted nylon 6, 6 during washing in an activated oxygen bleach-containing detergent. In the first part, fading observed in the UK-YO test and after repeated domestic laundering has been investigated in the case of ten acid leveling and six acid milling dyes.

The effect of after treatment with Intrafix PA and two-bath Fixogene A-XE and Fixogene C-XF was also studied. In the UK-TO test , acid leveling dyes, irrespective of whether an after treatment has been applied for not, exhibited a level of fading less than DELTA E = 4CIE LAB units at 50°C (i.e., a pass), but showed fading at 60°C. Meeting the pass criterion at 60°C depends on the depth of shade and the after treatment used. Under the same test conditions, acid milling dyes exhibited an acceptably low level of fading bath at 50°C and 60°C even in the absence of the after treatment. The after treatment offered some protection against fading in the UK-TO test.

This improvement was most noticeable fir acid leveling dyes when UK-TO test was conducted at 60°C. After multiple washing, acid leveling dyes exhibited unacceptably high levels of fading (< 10 units) when washed twenty times at 50°C, irrespective of whether the after treatment had been given or not, it concluded that the fading of selected acid dyes can be attributed to a wash down/ desorption effect, rather than an effect , rather than an oxidative bleach effect. In the second part, the same authors have attempted to correlate the fading exhibited by the same set of optionally treated sixteen acid dyes in both test methodologies, so as to have the most reliable means of predicting the level fading that an article may expect to experience after repeated laundering.

It is found that the best correlation was obtained between the fading of optionally after-treated acid dyes, when the UK-TO test was carried out at 60°C and the twenty machine washes were conducted at 40°C (correlation factor, 79%). The correlation factor increased to about 90% for after treated dyeings. The use of a preliminary wash test to eliminate low wash fastness acid dyes prior to conducting the UK-YO test did not improve the correlation between fading in the UK-TO test at 60°C and that after multiple machine washing at 40°C.

A pre-activated hydrogen/ peroxide activator system for bleaching and disinfection has been described by authors. In this study, different pre-activated solutions of hydrogen peroxide and TAED and acetyl caprolactum (ACL) were investigated with regard to the time-temperature stability of the produced peracetic acid (PAA). The PAA was formed under alkaline conditions and its pH value was reduced to 5.5 by adding citric acid at defined conditions.

The bleaching efficiency at 40°C and 60°C determined with a number of textile bleach test monitored, corresponded to that of the conventional PAA system. Under these acidic conditions, the formed PAA was sufficiently stable over a daily working period of 8hr. for same test monitors, the pre-activated solutions even outperformed the conventional one. With regard to disinfection, the performance of the pre-activated solutions was also comparable to that of the conventional product[20].

CATIONIC BLEACH ACTIVATORS FOR IMPROVING COTTON BLEACHING[21, 22]

Bleaching cotton with hydrogen peroxide requires either high temperatures or long dwell times. Recently developed cationic compounds that rapidly react with hydrogen peroxide to give peracids have been shown to substantially decrease the amount of time necessary to achieve a commercially acceptable white fabric. In this work, commercial recipes for cold pad batch bleaching of knitted cotton fabric were initially investigated. These recipes were used as benchmarks to assess the effectiveness of the cationic activator system. The cationic activator/hydrogen peroxide bleaching system was found to provide comparable fabric whiteness in pad batch processes at much shorter bleaching cycles, potential benefits include improved fabric physical properties and simpler bleach formulas.

Bleaching Method

Five cold pad-batch commercial recipes (A-E, Table I) recommended by commercial companies for cotton bleaching were applied to greige cotton, dwelled for different time intervals, washed using hot tap water (95C), and air dried. A similar cold pad-batch method was used with the cationic bleach activators, Unless otherwise specified, in the case of bleaching systems containing bleach activators, the bleaching bath was composed of caustic soda, wetting agent (2 g), chelator (1 g), $H_2 O_2$ (8g), and activator, the bath was diluted with tap water to 100 ml, and the pH was adjusted to 11.9. Different molar ratios of chelator-activator-$H_2 O_2$ were used. Padding was performed on a W. Mathis AG paddar to give 100% wet pickup. Fabric samples (20.0g) were padded twice and placed in sealed plastic bags at 25C.

Conclusions

In the presence of hydrogen peroxide an alkali, bleach activators generate peracids, which are more potent oxidants. In this study a cationic bleach activator was employed to assist a hydrogen peroxide hot bleaching system. It was observed that temperature had the greatest influence on the whiteness of cotton fabric, followed in turn by the bleach activator concentration, hydrogen peroxide concentration and time. The predicted and measured whiteness values from the optimized recipe for the cationic bleach activator were in close agreement. Confirming the validity of the quadratic model used.

By the addition of the cationic bleach activator to a conventional hot bleaching system it was possible to obtain a similar level of whiteness at lower temperature and reduced time while maintaining wettability. In regard to fibre damage after bleaching, the cationic bleach activator system gave less chemical damage than the conventional bleaching system, as indicated by the high residual degree of polymerization. This activator may therefore by beneficial when bleaching fibre blends containing delicate components, for example cotton/wool blends in which the wool is prone to damage in the hot alkaline conditions.

References:-

1. Shenai V. A., *"Technology of Bleaching and Mercerizing"*, Sevak Publications, (1996).

2. Sarvanan D., *"Journal of Textile Association"*, March-April (2005), pp277.

3. Gulrajani M. L., Sukumar N., *"Journal of Society of Dyers and Colourists"*, **100**(1), (1984),pp21.

4. Gulrajani M. L., *"Colourage"*, **36**(2), (1990) 19.

5. Ammayappan L., Muthukrishanan G., Sarvanan Prabhakar C., *"Man Made Textiles in India"* (1), **46**, (2003), pp23.

6. Pan N. C., Chatopadhay S. N., *"Indian Journal of Fibre and Textile Research"*, **29**, March(2004), pp79-84.

7. Prabaharan M., Venkatarao J., *"Indian Journal of Fibre and Textile Research"*, **28**, December(2003), pp437-443.

8. Moussa K., Traope, Gisela Buschle, *"Textile Chemist and Colourists and American Dyestuff Reporter"*, **32**, **12** (2000).

9. Nevin Cigdem Gurasy, Habip Dayioglu, *"Textile Research Journal"*, **73/4**(2003), pp297.

10. Yachmanev V. G., *"Textile Research Journal"*, 2001, pp527.

11. Chien Hualin., *"Textile Research Journal"*,2001, pp425.

12. Chakrabarti M., *"Indian Journal of Fibre and Textile Research"*, 1998, pp250.

13. Yang X. T., *"Textile Research Journal"*,2001, pp388.

14. Tzanov T., *"Colouration Technology"*, 2001, pp1.

15. Chen W., Chen D., *"Textile Research Journal"*,2001,pp441.

16. Khishigsuren A., *"Textile Research Journal"*,2001, pp487.

17. Gacen J., Caynda D., *"Journal of Society of Dyers and Colourists"*, 2000,pp13.

18. Phillips D., *"Journal of Society of Dyers and Colourists"*,2000, pp229.

19. Phillips D., *"Colouration Technology"*, 2001,pp148-152.

20. Shenai V. A., *"Colourage"*, **48/Suppliment**(2001), pp61-66.

21. Navin Cigben Gursoy., Ahmad El Shafai, *"AATCC Review"*, **4/8**(2004), pp37-40.

22. Sang Hoon Lim., Navin Cigben Gursoy., *"Colouration Technology"*, **120**(2004), pp118.

23. Milner Alam J., *"AATCC Review"*, **2/10**(2002), pp17-19.

24. Milner Alam J., *"International Dyer"*, April(2004), pp13-15.